Education
90

学以致用

Living What You Learn

Gunter Pauli

[比] 冈特·鲍利　著

[哥伦] 凯瑟琳娜·巴赫　绘

田　烁　王菁菁　译

上海远东出版社

丛书编委会

主　任：田成川

副主任：何家振　闫世东　林　玉

委　员：李原原　翟致信　靳增江　史国鹏　梁雅丽

　　　　任泽林　陈　卫　薛　梅　王　岢　郑循如

　　　　彭　勇　王梦雨

特别感谢以下热心人士对童书工作的支持：

匡志强　宋小华　解　东　厉　云　李　婧　庞英元

李　阳　刘　丹　冯家宝　熊彩虹　罗淑怡　旷　婉

杨　荣　刘学振　何圣霖　廖清州　谭燕宁　王　征

李　杰　韦小宏　欧　亮　陈强林　陈　果　寿颖慧

罗　佳　傅　俊　白永喆　戴　虹

目录

Contents

一个晴朗的夏日早晨，一只驯鹿正在湖边散步。她发现了一座半月形的建筑，便走近瞧了瞧。周围非常安静，一个人都没有，于是她决定去看一看建筑里面的样子。

A reindeer is taking a walk near the lake on a fine summer morning. She spots a half-moon shaped building and goes closer to have a look. It is very quiet and with no one around, she decides to peek inside.

一只驯鹿正在散步

A reindeer is taking a walk

为什么屋子里面还长着植物呢

Why are there plants growing indoors

"为什么屋子里面靠近天花板的地方还长着植物呢？"她自言自语道。

一个萨米小男孩听到了她的问题，回答道："这样做是为了保持空气清新，我们还会在教室里种一些蔬菜、水果和草药。"

"Why are there plants growing indoors, near the ceiling?" she wonders out loud.

A young Saami boy overhears her and replies: "We do that to keep the air clean, and we also grow some veggies, fruit and herbs inside our classroom."

"在室内而不是室外种植物，而且还是在教室里！真让我大开眼界！为什么不用温室呢？"

　　"我们学校的确有一个温室，但是在教室里面我们也需要植物。这样，不管是夏天还是冬天，教室里的空气都可以和苔原地区的一样又清新又干净了。"

"Growing plants under the roof and not in the ground, and that in a classroom! Now I have seen it all! Why not use a green house instead?"

"Our school does have a green house, but we need plants in the classroom so the air can be as fresh and clean as that of the tundra, in summer and winter."

为什么不用温室呢?

Why not use a green house instead?

一整天都只能呼吸着同样的空气

Breathe the same air all day long

"你说得没错，大家都需要新鲜的空气。露天而睡，可以呼吸大自然最好的空气！但是，我们的孩子们要在教室中度过那么长的时间，他们一整天都只能呼吸着同样的空气。"

"You are right, we all need fresh air. Sleeping under the open skies we breathe the best air there is! But when our children are spending so much time in a classroom, they all breathe the same air all day long."

"但是，他们不知道吗，污浊的空气也被关在室内不能疏散出去了。而且，一个孩子打喷嚏，其他孩子很快也会跟着生病。"

"嗯，不幸的是，每个地方都有这种情况发生。但是，我看到这所学校地下有通道，屋顶上有烟囱。"

"是的，这所学校建有地下通道，就像白蚁建造自己的巢穴那样。"

"But don't they know that the stale air is then trapped inside? And when one child sneezes, soon everyone will be sick?"

"Well, unfortunately that is what happens everywhere. But I see that this school has tunnels underground and chimneys on the roof."

"Yes, it is built with tunnels, like termites build their homes."

地下有通道，屋顶上有烟囱

Tunnels underground and chimneys on the roof

让自己家里保持凉爽

To keep their homes cool

"但是，白蚁这样做不是为了让自己家里保持凉爽吗？"驯鹿问道。

"这里也是同样的道理，就是为了让建筑变冷或变热。寒冷的空气通过地下通道从室外输送进来，在通道中，冷空气变暖。当冷空气进入建筑里时，温度就高多了。"男孩解释道。

"But don't termites do that to keep their homes cool?" asks the reindeer.

"The same logic is at work here, for cooling down and warming up buildings. The freezing cold air is channelled from outside through tunnels where it then warms up. By the time it gets into the building it is much warmer," explains the boy.

"这种做法很聪明。它说明白蚁的方法对于暖空气和冷空气都同样适用。我想知道，如果孩子们整天都能呼吸到新鲜空气，甚至冬天也是如此时，会发生什么？"

"That is smart. It means that the termite's logic works for both warm and cold air. What I want to know is what happens when children have fresh air all day, even in the winter?"

对于暖空气和冷空气

For both warm and cold air

更健康、更聪明

Healthier and smarter

"如你所知，如果所有的学校都这样做，保证充足的新鲜空气，那么孩子们在学业上就会表现得更好。如果我们教给他们这种方法，孩子们将会学以致用。"

"这确实能让我们这些生活在瑞典北部的孩子们更健康、更聪明。我们非常幸运，希望全世界所有的孩子都可以在自家附近有这样的学校。"

"You know, if all the schools do this, ensuring plenty of fresh air, then the kids will do better with their studies. And if we teach them about this, children will be living what they learn."

"It surely makes us kids living here in the North of Sweden healthier and smarter. We are very fortunate and I wish for all children around the world that they can attend schools like this one, close to home."

"如果更多的人想住到像这样的学校附近，那将会有更大的社区。我总是说，人越多越好！"

　　"我喜欢这样，"男孩回答道，"有更多的朋友可以一起玩耍，有更多的孩子可以互相学习，也有更多的人和我们一起庆祝生活的奇迹！"

　　……这仅仅是开始！……

"If more people want to come live near schools like this, there will be a bigger community. And the more the merrier, I always say!"

"I like that," responds the boy. "More friends to play with, more children to learn from and more people to celebrate the marvels of life with us!"

... AND IT HAS ONLY JUST BEGUN!...

······这仅仅是开始！······

... AND IT HAS ONLY JUST BEGUN! ...

Did You Know ?

你知道吗？

Poor quality indoor air is caused by small particles like dust, trapped gasses like CO_2, and micro-organisms like moulds and bacteria created or released by furniture, carpets and paints in the room.

糟糕的室内空气质量是由以下物质导致的：小颗粒物（如灰尘）、非流动气体（如二氧化碳）、微生物如室内家具、地毯、喷绘等产生或释放的霉菌和细菌。

在第三世界国家，室内空气污染是由燃烧木柴、木炭、粪便或农作物废料，用于取暖或做饭造成的。

In Third World homes indoor air pollution is caused by burning wood, charcoal, dung or crop waste for heating and cooking.

在发展中国家，室内空气污染可能是致命的，导致每年上百万人死亡。

二手烟严重影响同处一室的不吸烟者，让他们吸入了有毒气体，尤其是一氧化碳。空气动力学当量直径小于 2.5 微米的气体粒子（PM2.5）可以通过肺部的天然屏障，引发疾病。

Indoor plants remove volatile organic compounds such as benzene while releasing water and oxygen. Plants require hardly any energy and are as effective as industrial filtration in removing organic compounds.

室内植物可以去除挥发性有机物，比如苯，同时释放水分和氧气。这一过程中，植物几乎不需要消耗能量，却和工业过滤一样有效。

如果想保持室内空气新鲜、富含氧气，教室需要每小时通风三次。

If the indoor air quality is to be on par with fresh, oxygen-rich air then the total amount of air in the classroom should be replaced three times an hour.

3 次/小时

Moulds in the air trigger allergies and asthma. Insulation and other means of energy savings in homes such as sealed windows reduce air circulation and traps micro-organisms and particles inside buildings while it increases damp, which promotes the growth of micro-organisms.

空气中的霉菌会引发过敏和哮喘。家中的隔热设施和其他节能设备（如密封窗），会减少空气流通，让微生物和颗粒物留在建筑内，从而增加湿气，促进微生物的生长。

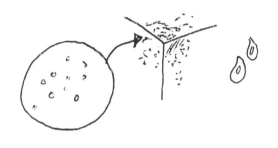

At the Laggarberg School in Timrå (Sweden) the total volume of air in all classrooms is replaced with fresh air every two hours. The more students present in the sports hall, the more often the air is replaced with fresh air. Children attending this school remain healthy, have an increased attention span and study better.

在瑞典蒂姆罗的拉格堡学校，所有教室每两小时就会彻底通风一次。体育馆里的学生越多，通风次数就越频繁。这所学校的孩子一直都很健康，注意力更集中，学习更好。

你想去那种到处都是隔热板和密封窗的学校上学吗?

Would you like to attend a school with a lot of insulation and windows that do not open?

Does anyone in your family smoke, subjecting you to the effects of second hand smoke? What do you say to them?

你家里有人吸烟让你遭受二手烟的影响吗? 你会和他们说些什么?

在屋顶上安上烟囱, 让上升的热空气散发出去, 这样做有道理吗?

Does it make sense to have chimneys on the roof to let out hot air that rises?

Is it possible to have good health and energy efficiency at the same time?

提高能效的同时也保持身体健康, 这有可能实现吗?

Does anyone around you smoke? Second-hand smoke is a serious health hazard because the particles released by burning cigarettes are so small that they pass through the natural defence barriers of the lungs. Draw up a list of arguments why you should be protected against these toxins. Offer simple, logical and convincing arguments. Then take the position of a smoker, and argue why that person believes he or she should be allowed to smoke in public places, even if it exposes others to health risks. Compare the two arguments and draw your conclusions.

你周围有人吸烟吗？二手烟对身体健康的危害非常大，因为香烟燃烧释放出来的颗粒物十分微小，能够通过肺部的天然屏障。写出你为什么要抵制这些有害物质，提出一些简明、有逻辑、令人信服的理由。然后，从一个吸烟者的角度，想一想人们凭什么冒着健康风险允许他在公共场所吸烟。对比一下这两种情况，给出你自己的结论。

学科知识
Academic Knowledge

生物学	驯鹿的生态系统和远足习性；空气过滤植物与微生物在生长介质中相互作用，共同净化空气，芦荟、棕榈树、蕨类植物、常春藤、菊花、兰花、百合花是最好的过滤植物；缺氧会导致大脑反应迟缓，身体机能降低，包括注意力降低、记忆力变差。
化 学	植物能去除空气中的甲醛、苯、甲苯、二甲苯和三氯乙烯；氡是某些建筑材料中释放的一种放射性原子气体，是影响现代建筑物室内空气质量的最普遍隐患。
物 理	将空气过滤植物放在靠近天花板的位置效果最好，这样可以让氧气下沉；血液中含氧量下降会引发肺动脉和肺静脉收缩，迫使心脏跳动更加剧烈，呼吸频率增加。
工程学	自然通风的设计要求利用低气压将含有害物质的空气输送到植物走廊中，从而得到净化；瑞典的地下通道利用地热取暖，加热了户外的冷空气；瑞典建筑规范对自然光和自然通风有相应的标准。
经济学	室内空气质量决定了办公室员工的生产力和学校学生的学习能力；当人们搬到教学质量更好的学校附近居住，房屋需求的增加将导致土地升值，这意味着良好的教育设施可以提高土地价值。
伦理学	如果学校或办公室节省了能源成本，却对空气质量产生不利影响，这可以接受吗？
历 史	玻璃窗是埃及亚历山大市的首创，在罗马时代（公元100年）已经开始使用了；英国在17世纪广泛使用玻璃窗，16世纪早期，英国玻璃和铸铁产量的激增导致了大规模的森林砍伐活动；中国和日本曾经使用纸窗。
地 理	萨米人又称拉普人，是北极地区的原住人口；冻土带的气候；瑞典北部的年轻一代正迁往南部，导致这一区域适龄入学儿童的数量减少，学校越来越少。
数 学	计算管道直径、烟囱数量、气压和温度，以及每小时通风所需的空气总量，以保持高含氧量和低二氧化碳水平。
生活方式	城市中久坐不动的生活方式越来越普遍，让我们不能享受海滨、森林的新鲜空气，降低了我们的免疫力。
社会学	人们搬到离学校近的地方居住；1万多年来，萨米人（拉普人）的生活依靠驯鹿的鹿肉、鹿奶和鹿皮，但是，因为开发滑雪胜地，驯鹿的迁徙路线受到了影响。
心理学	呼吸新鲜空气可以减轻压力，使精力充沛，增加快乐和幸福感。
系统论	对能效的追求产生了负面效应，即病态建筑物综合征；创新的建筑设计能兼顾能源效率和我们的身体健康。

情感智慧
Emotional Intelligence

驯　鹿

驯鹿有求知欲，想知道教室的天花板附近为什么还长着植物，并提出了温室种植的建议。她很喜欢教室里通过自然循环而获得的新鲜空气，意识到要在新鲜空气和节能措施之间做好平衡。驯鹿有敏锐的观察力，发现了更多关于如何使用当地现有资源解决新鲜空气供给问题的案例。从萨米男孩分享的信息中，她学到了一些让孩子们保持健康、集中精力、学习更好的普遍而有智慧的方法。针对人们搬往有好学校的地方居住，让那些区域炙手可热的现象，她分享了自己的见解。

萨米男孩

萨米男孩乐于分享自己的见解，表明了不管冬夏，教室里的空气都应该和苔原地区一样清新的观点。他质疑环境，并将供氧能力的积极作用和孩子们的健康状况联系在一起。他理解"白蚁筑巢技术"，完全掌握了它的运行原理，也听懂了驯鹿关于好学校与大社区紧密联系的观点。萨米男孩看到了更加光明的未来，并愿和幸福、健康、聪明的孩子们一起分享。

艺术
The Arts

用数学公式解释地下通道和烟囱如何让空气升温、冷却，在教室天花板附近种植物来净化空气，这些都不是容易的事。画出你想象中的空气流通的学校是什么样子的，向其他人展示你的作品，帮助他们理解这种建筑设计如何让孩子们能呼吸足够的氧气，让他们不仅学习更好，还能保持健康快乐。

思维拓展
Systems: Making the Connections

在提高能效、减少化石燃料消耗的实践中，建筑技术发生了转变，建筑的密封性更强了，空气和水分都留在了室内。虽然这有助于实现较低的能源消耗，但也有不利影响，比如过高的湿度会刺激霉菌、动物皮屑（死皮）、植物花粉和微生物的增长，这些物质都有可能成为过敏原。使用地毯又进一步增加了空气中颗粒物的含量。在学校、家庭和办公室使用的固体、液体化合物，会排放出大量的气态化学物质。这些挥发性有机化合物过去通常可以通过开窗或过滤等方式去除，但是现在大部分的建筑设计没有可以控制开关的窗户，也可通过精密过滤的方式去除挥发性有机化合物，但过滤器成本很高，这样就有了引发病态建筑物综合征的风险。然而，让现代建筑设计实现健康、多氧与节能并存是可能的。这要求一种与众不同、有创新思维的建筑设计，它以几个世纪以来广泛应用的设计为基础，已经被全世界众多建筑师广泛提及。他们已经证实，创新的建筑设计不仅能减少建筑的资金成本，还能减少建筑的运行成本。其效益不仅体现在货币上，还包括提高了劳动生产率，而最主要的效益则是保持身体健康。当通风技术与植物精准种植技术相结合时，效益就会实现前所未有的增加。瑞典蒂姆罗的拉格堡学校提供了一个实现这一模式的优秀案例——利用植物来净化整个建筑里的空气，当体育馆里的人多起来时，室内的空气会变得更清新。额外的效益则是，在这所学校学习的孩子们学到了更多生物多样性的奇观，对自然界中植物所扮演的角色有了更加积极的理解。

动手能力
Capacity to Implement

相比于打地下通道、安装烟囱来优化空气，在家里摆放植物的方案要容易得多，不仅能美化环境，还有利于身体健康。和你的父母讨论一下，在家里多摆放一些植物。思考一下哪种植物最合适，以及每个房间需要多少植物。使用LED灯，晚上的光线会欺骗植物，让它们误以为太阳光还在，并继续生长。先从你自己的卧室开始，然后再给家里的其他房间也摆上植物吧！

故事灵感来自
This Fable Is Inspired by

安德斯·尼奎斯特
Anders Nyquist

安德斯·尼奎斯特在瑞典北部出生和成长。他以建筑师身份毕业，专攻建筑历史与考古学。自1966年起，安德斯·尼奎斯特在健康与生态建筑设计方面进行了开拓性工作，并赢得了很多奖项。1990年，在从事传统建筑工程职业生涯之后，他与妻子英格丽和女儿凯伦创办了自己的设计工作室。他致力于利用建筑和设计创建社区，以刺激当地经济发展，并推动创新。拉格堡学校的设计是世界范围内的建筑标杆，引起了全球许多艺术家、政治领袖的兴趣。这所学校的成功运行促进了瑞典于默奥"绿色地带"的发展，这是一个成功的案例，展现了工业区是如何模仿自然生态系统中物质、水和能量流动的。

图书在版编目（CIP）数据

冈特生态童书.第三辑修订版:全36册:汉英对照 /
(比)冈特·鲍利著;(哥伦)凯瑟琳娜·巴赫绘;
何家振等译.—上海:上海远东出版社,2022
书名原文:Gunter's Fables
ISBN 978-7-5476-1850-9

Ⅰ.①冈… Ⅱ.①冈… ②凯… ③何… Ⅲ.①生态环
境–环境保护–儿童读物—汉、英 Ⅳ.①X171.1-49

中国版本图书馆CIP数据核字(2022)第163904号
著作权合同登记号图字09-2022-0637号

策　　划　张　蓉
责任编辑　程云琦
封面设计　魏　来李　廉

冈特生态童书
学以致用
[比]冈特·鲍利　著
[哥伦]凯瑟琳娜·巴赫　绘
田　烁　王菁菁　译

记得要和身边的小朋友分享环保知识哦！
八喜冰淇淋祝你成为环保小使者！